Sea Turtles

Young Explorer Series: Book 2

ISBN 978-1-7352616-1-4

Humble Bee Press LLC

www.facebook.com/humblebeepress

Dedication

For my grandson Braxton. Your thirst for learning has played a big role in my *Young Explorer Series*. I truly hope you never stop asking your favorite question . . . "Why?"

Love you always,

Grandma

SEA TURTLES

Sea turtles and land turtles look similar, but sea turtles have flippers instead of legs. These flippers help them to gracefully swim about in their underwater habitat.

Unlike land turtles, sea turtles cannot tuck their head or flippers back into their shells. This makes sea turtles vulnerable to ocean predators and entanglement of marine debris.

A Sea Turtle

Land Turtle

Sea turtles spend their whole life at sea, except for adult females who return to shore to nest and lay eggs. Normally, in the dark of night, the mother will dig her nest and lay anywhere between 80-160 eggs.

When she is finished, she returns to sea and the eggs will develop on their own.

Depending on the species, the eggs will hatch anywhere between 45-90 days. After hatching, the babies wait a few days, while gaining strength to be able to climb from their nest and start their journey to the sea.

Watching "hatchlings" as they struggle to climb from their nest and instinctively find their way to the water is always an amazing sight.

Everything from holes in the sand to driftwood and crabs can be obstacles along the way. Predators like birds, raccoons, and fish are another danger for these vulnerable creatures; some experts estimate only 1 out of 1000 hatchlings will survive to adulthood under natural conditions.

There are seven different types of sea turtles.

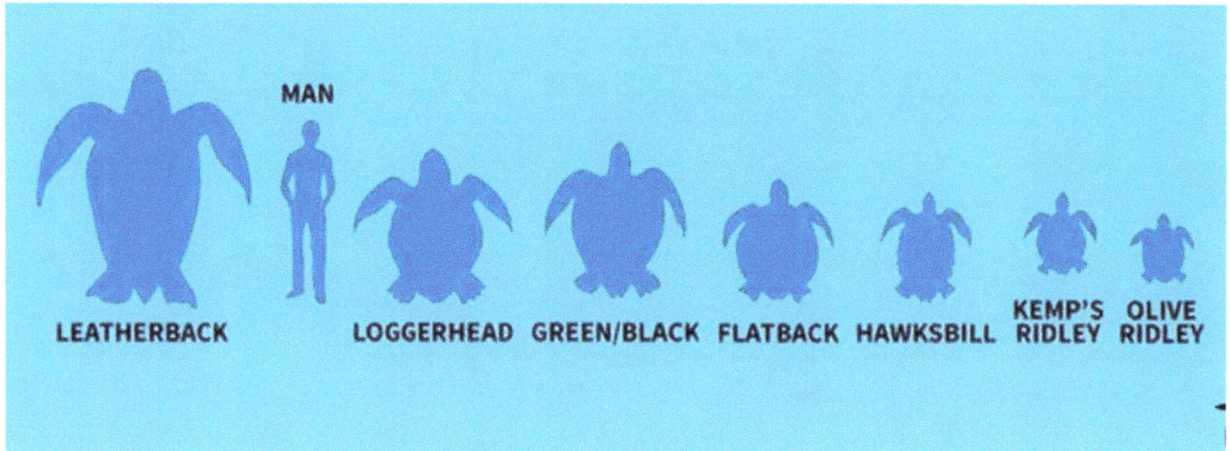

Leatherback

Loggerhead

Green

Flatback

Hawksbill

Kemp's Ridley

Olive Ridley

Characteristics

The seven species can be identified by their size, colors, and carapace shapes.

KEMP'S RIDLEY

The Kemp's ridley is the smallest of the sea turtle species. The Kemp's ridley has a triangular-shaped head with a slightly hooked beak.

Kemp's ridley are mainly found in the Gulf of Mexico and up through the eastern Atlantic seaboard.

OLIVE RIDLEY

The olive ridley gets its name from the olive-green color of its heart-shaped carapace (top shell).

The species is one of the smallest of the world's sea turtles and is found primarily in the tropical regions of the Pacific, Indian, and Atlantic Oceans.

LOGGERHEAD

The loggerhead is named for its massive head and strong jaws. Their jaws allow them to crush conchs, bivalves, and horseshoe crabs. They are mainly carnivores (meat eaters). They also eat jellyfish, shrimp, sponges, fish, and sometimes even seaweed and sargassum.

They are the largest hard-shell turtle (leatherbacks are bigger, but they have soft shells).

The average weight of a loggerhead is 250 pounds, although loggerheads weighing more than 1000 pounds have been found.

LEATHERBACK

The leatherback is the largest turtle in the world. They can reach up to 2200 pounds. Named for their tough rubbery skin, they are the only species of sea turtle that does not have a hard shell.

Leatherbacks have been tracked swimming over 10,000 miles a year between nesting and foraging grounds. They are also accomplished divers with the deepest recorded dive reaching nearly 4,000 feet—deeper than most marine mammals.

GREEN

The green turtle is one of the largest hard-shelled sea turtles. They are the only sea turtle that, as an adult, is strictly herbivorous (diet of plant materials). The diet is what gives their cartilage and fat a greenish color.

Green turtles live all over the world mainly near the coastline and around islands in bays and protected shores. Rarely are they observed in the open ocean.

HAWKSBILL

The hawksbill is named for its narrow head and hawk-like beak. Their carapace is orange, brown or yellow. It is one of the smaller sea turtles.

Hawksbill are found in tropical and subtropical waters of the Atlantic, Pacific and Indian Oceans.

The female nests between 3 to 6 times per season and lays an average 160 eggs in each nest. The eggs will hatch in about 60 days.

FLATBACK

The flatback is named for its very flat shell. Its carapace is olive-grey with brown/yellow tones.

They eat sea cucumbers, jellyfish, mollusks, prawns, bryozoans, other invertebrates, and seaweed.

Flatbacks are found only in the waters around Australia and Papua, New Guinea in the Pacific.

FUN FACTS

⚓ Sea turtles breath air. They can hold their breath up to 5 hours under water.

⚓ Sea turtles have ears. Not visible on the outside of their head, their ears allow them to hear vibrations in the water.

⚓ Many ocean organisms use sea turtle shells as a home or a feeding station.

⚓ Female sea turtles migrate thousands of miles, sometimes to the same beach where they were born, in order to nest and lay eggs.

⚓ The gender of a sea turtle is determined by the temperature of the developing eggs. Warmer temperatures produce females and cooler temperatures produce males.

⚓ Once hatchlings leave the nest, they are often not seen again until they return to coastal waters, sometimes 10 years later. This is known as the "lost years."

⚓ Sea turtles are the oldest living animal on the planet. Dating back 110 million years, they were here even before dinosaurs.

⚓ All seven species of sea turtles have been listed as either threatened or endangered according to World Wildlife Foundation (WWF).

Ocean Plastic & Sea Turtles

As you have read, sea turtles have plenty of natural obstacles to make it to adulthood, but there are also man-made ones. One of the biggest man-made threats is plastic.

Sea turtles are affected by plastic during every stage of their life. They crawl through plastic on the way to the ocean as hatchlings, swim through it while migrating, confuse it for jellyfish (one of their favorite foods), and then crawl back through it as adults.

Hundreds of thousands of sea turtles, whales, and other marine mammals, and more than 1 million seabirds die each year from ocean pollution and ingestion of plastic.

How Can We Help?

- Stop using plastic straws, even in restaurants. If a straw is a must, purchase a reusable stainless steel or a glass straw.

- Use a reusable produce bag. A single plastic bag can take 1,000 years to degrade.

- Buy boxes instead of bottles. Cardboard is more easily recycled than plastic.

- Reuse containers for storing leftovers or shopping in bulk.

- Use a reusable bottle or mug for your beverages, even when ordering from a to-go shop.

- Bring your own container for take-out or your restaurant doggy-bag since many restaurants use styrofoam.

- Use matches instead of disposable plastic lighters or invest in a refillable metal lighter.

- Don't use plasticware at home and be sure to request restaurants do not pack them in your take-out box.

- Stop releasing balloons for celebrations. Remember, those balloons have to land somewhere.

A Word from the Author

Our majestic sea turtle is the oldest living animal on this planet. These docile and graceful creatures have captured the hearts of many.

The fact that *we* are such a big contributor to their endangered status is heartbreaking.

Until recently, many were unaware of how our use of plastics were affecting marine life. I believe bringing awareness is the first step in helping to preserve all marine life.

"I did then what I knew how to do. Now that I know better, I do better." **Maya Angelou**

For more information on the Young Explorer Series visit
www.facebook.com/humblebeepress